John Minot Rice, William Woolsey Johnson

On a New Method of Obtaining the Differentials of Functions with Especial Reference to the Newtonian Conception of Rates or Velocities

John Minot Rice, William Woolsey Johnson

On a New Method of Obtaining the Differentials of Functions with Especial Reference to the Newtonian Conception of Rates or Velocities

ISBN/EAN: 9783337811815

Printed in Europe, USA, Canada, Australia, Japan

Cover: Foto ©berggeist007 / pixelio.de

More available books at **www.hansebooks.com**

ON

A NEW METHOD

OF OBTAINING THE

DIFFERENTIALS OF FUNCTIONS

WITH ESPECIAL REFERENCE TO THE

NEWTONIAN CONCEPTION OF RATES OR VELOCITIES

BY

J. MINOT RICE

PROFESSOR OF MATHEMATICS IN THE UNITED STATES NAVY

AND

W. WOOLSEY JOHNSON

PROFESSOR OF MATHEMATICS IN SAINT JOHN'S COLLEGE ANNAPOLIS MARYLAND

REVISED EDITION

NEW YORK

D. VAN NOSTRAND PUBLISHER

23 MURRAY STREET AND 27 WARREN STREET

1875

THIS pamphlet is a revised edition of a paper which was read before the American Academy of Arts and Sciences, January 14, 1873, and subsequently published in its Proceedings.

It is now reproduced for the purpose—first, of presenting a new method of deriving the Differentials of Functions by means of their Algebraic characteristics with the aid of a few elementary properties easily established; and, secondly, of showing that the method of rates or fluxions may be advantageously used for the purposes of instruction, and the use of infinitesimals, limits and series entirely avoided until the student is well grounded in the elements of the Calculus; thus securing the advantages afforded by the real and precise definitions of this method, instead of sacrificing them by employing the difficult and readily misconceived notions of limits or infinitesimals in deducing the formulas for differentiation.

In the original paper it was shown that the new method of deducing the differentials was applicable to all the functions of a single variable; some of these applications are now omitted to be replaced by other methods which we consider preferable for the purposes of elementary instruction.

It is our intention to publish a text-book prepared in accordance with this plan, the first part of which has already been printed for the use of the cadets at the U. S. Naval Academy.

J. M. R.
W. W. J.

ANNAPOLIS, MD., *July*, 1875.

CONTENTS.

6 CONTENTS.

THE NEWTONIAN METHOD OF FLUXIONS.

1. It is a well-known fact that writers on the Differential Calculus deduce the same elementary theorems from fundamental conceptions which have little or no apparent resemblance, and frequently employ methods which present to the conscientious student difficulties of a character too formidable to be ignored.

Notwithstanding the unusual attention which many of the ablest mathematical writers since the time of Newton and Leibnitz have bestowed upon this subject it is undoubtedly true that many instructors still tacitly permit their students to follow the trite precept of D'Alembert,—"*Allez en avant, et la foi vous viendra.*" Few habits are more pernicious to the student of mathematics than that of following rules founded upon principles which he does not thoroughly comprehend; yet it is precisely this habit that D'Alembert's precept tends to confirm. Faith comes—only too soon.

In the words of Condillac—" Ce n'est point par la routine qu'on s'instruit, c'est par sa propre réflexion; et il est essential de contracter l'habitude de se rendre raison de ce qu'on fait; cette habitude s'acquiert plus facilement qu'on ne pense; et une fois acquise, elle ne se perd plus."

. 2. Most of the French writers on the Calculus have adopted a method of treating the subject which may be characterized as a combination of the method of limits with that of infinitesimals. One of the best examples is the work of J. A. Serret, Paris, 1868. This method, although excellent in extensive treatises like those of Serret and Bertrand, seems to us far too difficult for a beginner not possessed of unusual mathematical ability, especially as it involves several fundamental propositions very hard to comprehend. In nearly all cases, it will be found best for the student not to attempt these works until he has prepared himself by studying one of a less formidable character.

3. A distinguished English writer of mathematical text-books (Mr. I. Todhunter), who has himself adopted the method of limits, remarks that—" A difficulty of a more serious kind which is connected with the notion of a limit, appears to embarrass many students of this subject, namely, a suspicion that the methods employed are only approximative, and therefore a doubt as to whether the results are absolutely true. This objection is certainly very natural, but at the same time by no means easy to meet, on account of the inability of the reader to point out any definite place at which his uncertainty commences." *   This remark seems to us not only to go to the root of the difficulty, but also to suggest an excellent mode of testing the student's comprehension of the subject when taught by the ordinary methods.

---

* *Todhunter's Differential Calculus*, Macmillan and Co. London, 1860, page 12.

4. Our plan is to return to the method of fluxions, and making use of the precise and easily comprehended definitions of Newton, to deduce the formulas of the Differential Calculus by a method which is not open to the objections which were largely instrumental in causing this view of the subject to be abandoned; and one which is not in the slightest degree suggestive of approximation.

Inasmuch as the method of fluxions is now nearly obsolete, we would refer the reader, who may desire a brief and lucid statement of its principles and peculiarities, to *Montucla's History of Mathematics*, Volume II, pages 320, 321, 322, and 323 (the edition before us is that of 1758).

5. The objections hitherto urged against this view of the subject may be divided into two classes; first, those which refer to the mode of deducing the formulas, and secondly, objections to the definitions employed. The following extract from Lacroix's treatise will serve as an example of the first class of objections.

6. "Newton supposa les lignes engendrées par le mouvement d'un point, et les surfaces par celui d'une ligne; et il appela fluxions les vîtesses qui réglaient ces mouvemens. Ces notions, quoique très-rigoureuses, sont étrangères à la Géométrie, et leur application peut être difficile. Il est bien vrai qu'en imaginant un point qui se meuve sur une ligne, pendant qu'elle est emportée parallèlement à elle-même, avec une vitesse uniforme, on peut représenter une courbe quelconque; mais la

vitesse du point décrivant étant variable à chaque instant, *on ne peut la déterminer qu'en recourant soit à la méthode des Anciens ou d'exhaustion, soit à celle des premières et dernières raisons*, et c'est presque toujours de celle-ci que Newton s'est servi; ensorte que les fluxions n'etaient, à proprement parler pour lui, qu'un moyen de donner un object sensible aux quantités sur lesquelles il opérait."*

The italics in the above extract are ours. It will be observed that the clause thus distinguished contains the point of M. Lacroix's objection; an examination of the following pages will show that it does not apply to our method.

7. The two following extracts contain substantially all the objections of the second class that we have thus far met with, and also replies to them which seem to us to be sufficiently conclusive.

† " Cette observation a été le prétexte d'une objection élevée contre la méthode des fluxions; car, a-t-on dit, c'est introduire dans la Géométrie qui appartient aux Mathématiques pures, la notion des vitesses qui n'appartient qu'aux Mathématiques mixtes, et définir une idée qui doit être simple, par une autre qui est complexe.

---

* *Traité du Calcul Differentiel et du Calcul Intégral.*   Par S. F. Lacroix.   Seconde édition.   Tome premier.   Paris, 1810.   Préface, pages **xv** et **xvi**.

† *Carnot—Réflexions sur la Métaphysique du Calcul Infinitesimal.* Quatrième édition.   Paris, 1860.   Pages 114 et 115.

" Mais cette objection est assez frivole : car la véritable chose à considérer est de savoir si la théorie est plus facile à saisir de cette manière que d'une autre. Le classement que nous faisons des sciences est assez arbitraire. Nous plaçons la Géométrie avant la Mécanique dans l'ordre de la simplicité, mais les parties transcendantes de la première sont bien plus abstraites que les parties élémentaires de la seconde, et, comme le dit Lagrange, 'chacun a ou croit avoir une idée nette de la vitesse'; ce n'est donc pas prendre une marche contraire à l'esprit des Mathématiques, que de définir les fluxions par les vitesses."

* " On a reproché à Newton de faire intervenir, sans nécessité, dans ce mode d'exposition, la notion du temps et celle du mouvement. Le reproche peut être fondé, quant à la notion du mouvement, à laquelle rien n'oblige, en effet, de recourir; mais on devait remarquer que la notion du temps intervient ici par la nature des choses, en raison de ce que le temps est la seule variable essentiellement indépendante, et la seule dont la variation soit essentiellement uniforme, ou la fluxion constante.

" Dans tous les cas, la conception de Newton, appliquée aux grandeurs qui varient effectivement avec le temps, a l'avantage de fixer la signification *réelle* des fonctions dérivées, et par là même de donner à l'avance la raison du rôle qu'elles jouent dans les applications de l'analyse à la discussion des phénomènes physiques. Newton se proposait aussi de fonder la théorie

---

* *Cournot—Théorie des Fonctions et du Calcul Infinitesimal.* Paris, 1841.

des fonctions sur une idée que l'esprit pût saisir directement, sans passer par la considération des limites et sans s'assujettir à une marche jusqu'à un certain point détournée et indirecte. Il entendait exprimer directement la continuité dans la variation des grandeurs, au moyen du phénomène le plus familier où cette continuité tombe sous les sens. On a objecté, d'après d'Alembert, que, pour *définir* une vitesse continuellement variable, il faut toujours recourir à la considération des limites; mais, en faisant cette objection, on a mal à propos subordonné la précision des idées à leur définition logique. Un concept existe dans l'entendement, indépendamment de la définition qu'on en donne ; et souvent l'idée la plus simple dans l'entendement ne comporte qu'une définition compliquée, quand elle n'échappe pas à toute définition. Tout le monde a une idée directe et exacte de la similitude de deux corps, quoique peu de gens puissent entendre les définitions compliquées que les géomètres ont données de la similitude."

The reader is desired to notice particularly the positive advantages of the Newtonian method which M. Cournot has so clearly and forcibly set forth in the second paragraph of the above extract.

PROPOSED METHOD OF TREATING THE DIFFERENTIAL CALCULUS.

8. When a quantity varies uniformly, the constant numerical measure of its rate is the increment received in the unit of time. When, however, the variation is not uniform, we would *define the numerical measure of the rate at any instant as the increment which would be received in a unit of time, if the rate remained uniform from and after the given instant.*

This definition corresponds with the usage of Mechanics, in accordance with which a body moving with a variable velocity is said to have at a given instant a velocity which would carry it thirty-two feet in one second.

9. To avoid departing too much from well-established usage, the term *differential* will be frequently used in this paper instead of *rate.*

The *rate* or *differential* of x will be denoted by dx, and that of $f(x)$ by $d[f(x)]$ (these symbols being always used by us to denote finite quantities.)

The rate of the independent variable, or the value of dx, is regarded as arbitrary in the same sense in which the value of the variable x is itself arbitrary.

Thus, particular values of these two quantities may constitute the datà of a question like the following: What is the value of $d(x^2)$ when x has the value 10, and dx the value 4?

To differentiate a function of x is to express $d[f(x)]$ in terms of x and dx in such a manner as to furnish a general formula by which $d[f(x)]$ may be computed for any given values of x and of dx.

ELEMENTARY PROPOSITIONS.

10. The following propositions are immediate deductions from the above method of measuring rates :—

I. *The Differential of* x + h.

Since any simultaneous increments of x and of $x + h$ must be identical, the increments which would be received by each, if they continued to vary uniformly with the rates denoted by dx and $d(x + h)$, are equal. Hence the rates are equal, or

$$d(x+h) = dx.$$

II. *The Differential of* x + y.

Since any increment of $x + y$ is the sum of the simultaneous increments of x and of y, the same relation exists between the increments which would be received if x and y (and consequently $x + y$) continued to vary uniformly with the rates denoted by dx, dy, and $d(x + y)$. Hence

$$d(x + y) = dx + dy.$$

III. *The Differential of* mx.

Since any increment of mx must be m times the corresponding increment of x, the same relation must exist between the increments which would be received if x (and consequently mx) continued to vary uniformly with the rates denoted by dx and $d(mx)$. Therefore

$$d(mx) = m\,dx.$$

THE RATIO OF THE RATES OF A VARIABLE AND ITS FUNCTION.

11. Let y denote a linear function of x, that is

$$y = mx + b. \dots\dots\dots\dots\dots\dots(1)$$

By propositions I and III,

$$dy = m\,dx,$$

or

$$\frac{dy}{dx} = m \dots\dots\dots\dots\dots\dots(2)$$

In this case, the ratio of the rate or differential of the function to that of the independent variable is constant, its value being independent, not only of x, but also of dx. Thus, if we give to dx any arbitrary value, it is evident from equation (2), that dy must take a corresponding value such that the ratio of these quantities shall always retain the constant value m. Assuming rectangular co-ordinate axes, if y be made the ordinate corresponding to x as an abscissa, the point (x, y) will, as x varies, generate a straight line. The direction of the motion of the point is constant, and depends upon the value of m. Since $\frac{dy}{dx}$ is equal to m, it is the trigonometrical tangent of the constant inclination of the direction of the generating point to the axis of x.

When y is not a linear function of x, the direction of the motion of the generating point is variable, and consequently the value of $\frac{dy}{dx}$ is variable.

12. Making now the arbitrary quantity dx a constant, dy will be a variable. Suppose that, the generatrix having arrived at a given point, the ordinate y continues to vary uniformly with the rate denoted by dy at the given point; the value of $\frac{dy}{dx}$ will then become constant. The generatrix will now continue to move uniformly in the direction of the curve at the given point, and therefore the value which $\frac{dy}{dx}$ has at this point is that of the trigonometrical tangent of the inclination of the curve to the axis of x at this point. The line *now* described by the generatrix is called a tangent line to the curve, in accordance with the following general definition: *The tangent line to a curve at a given point is the line passing through the point, and having the direction of the curve at that point.*

IV. *The Ratio of the Rates is independent of their absolute values.*

13. Since the direction of the curve (or of the tangent line) at the point having a given abscissa is determined by the *form* of the function, or equation to the curve, the value of $\frac{dy}{dx}$, which is the trigonometrical tangent of the inclination of this direction, must be independent of the arbitrary quantity dx, which merely determines the velocity of the generating point.

In general, the value of $\frac{d[f(x)]}{dx}$ will change with that of x; $\frac{d[f(x)]}{dx}$ is, therefore, independent of dx, but is generally a function of x.

$d[f(x)]$, when expressed in terms of x and dx, is of the form

$$d[f(x)] = f'(x).dx$$

in which $f'(x)$ is another function of x.

In the ordinary methods the introduction of an equivalent proposition is, for the most part, avoided, by rejecting from the ultimate value of $\Delta [f(x)]$ all terms containing powers of Δx higher than the first.

14. We shall now proceed to show how, from the four elementary propositions, the differentials of the functions both algebraic and transcendental may be deduced. These propositions are here recapitulated for convenience of reference :—

I. $d(x+h)=dx.$

II. $d(x+y)=dx+dy.$

III. $d(mx)=mdx.$

IV. $\frac{d[f,x]}{dx}=f'(x)$, a function of x, independent of dx.

ALGEBRAIC FUNCTIONS.

THE DIFFERENTIAL OF THE SQUARE.

15. In establishing the formulas for the differentiation of the simple algebraic functions of an independent variable, we find it most convenient to begin with the square.

We first deduce a relation between two values of the derivative of the function, and the corresponding values of the independent variable; for this purpose we assume two values of the variable having a constant ratio m, thus :

$$z = mx, \text{ then will } dz = mdx \dots\dots\dots\dots\dots(1)$$

and
$$z' = m'x', \quad \therefore \quad d(z') = m'd(x') \dots\dots\dots\dots(2)$$

Dividing equation (2) by equation (1),

$$\frac{d(z^2)}{uz} = m\frac{d(x^2)}{dx}, \quad \dots\dots\dots\dots\dots(3)$$

a relation between two values of the derivative; putting for m its value $\frac{z}{x}$ we obtain

$$\frac{d(z^2)}{dz} = \frac{z}{x} \cdot \frac{d(x^2)}{dx}, \dots\dots\dots\dots\dots(4)$$

the relation required. Dividing by z to *separate* the variables, we have

$$\frac{1}{z} \cdot \frac{d(z^2)}{dz} = \frac{1}{x} \cdot \frac{d(x^2)}{dx} \quad \dots\dots \dots\dots\dots(5)$$

This equation is true for all values of the quantities x, z, dx and dz; for, by Theorem IV, $\frac{d(z^2)}{dz}$ is independent of dz, and $\frac{d(x^2)}{dx}$ is independent of dx, the derivatives being functions respectively of z and of x simply; moreover, the equality exists independently of any particular value of m, since it has been eliminated.

The first member of this equation, being independent of the value of dz, can be a function of no variable quantity except z, and so likewise the second member can be a function of no variable quantity except x.

If, therefore, we denote x^2 by $f(x)$, and adopt the usual notation for the derivative, we may write equation (5) thus:—

$$\frac{1}{z} \cdot f'(z) = \frac{1}{x} \cdot f'(x) \quad \dots \dots \dots \dots (5')$$

Now, since the form of the first member of this equation is the same as that of the second member, it follows that the value of the expression $\frac{1}{x} \cdot f'(x)$ or $\frac{1}{x} \cdot \frac{d(x^2)}{dx}$ does not change when x is changed to z, the latter representing, it must be remembered, *any other* value of the independent variable; the value of this expression is therefore constant, and denoting it by c, we write

$$\frac{1}{x} \cdot \frac{d(x^2)}{dx} = c; \quad \dots \dots \dots \dots \dots (6)$$

whence

$$d(x^2) = cx\,dx \quad \dots \dots \dots \dots \dots (7)$$

To determine the unknown constant c we apply this result to the identity

$$(x+h)^2 = x^2 + 2hx + h^2; \quad \dots \dots \dots \dots (8)$$

differentiating each member by formula (7), we have

$$c(x+h)d(x+h) = cx\,dx + 2h\,dx; \quad \dots \dots \dots \dots (9)$$

since $d(x+h)=dx$, equation (9) reduces to

$$chdx=2hdx,$$

or

$$(c-2)hdx=0\ ;$$

and, since h and dx are arbitrary quantities, we have

$$c = 2,$$

which substituted in (7) gives

$$d(x^2) = 2xdx.$$

16. The process of which the above is an example may be applied in the case of all those functions whose differentials it is desired to deduce independently. Although it is somewhat elaborately explained in the above application, we add the following more general description of the method pursued.

We assume a new variable z, connected with x by a relation admitting of a comparison of dz and dx, and at the same time such that $d[f(z)]$ and $d[f(x)]$ may likewise be compared; in other words, such that the relation between z and x, and also that between $f(z)$ and $f(x)$ can be differentiated without the introduction of unknown differentials except those denoted by $d[f(z)]$ and $d[f(x)]$.

By division, the ratios $\frac{d[f(x)]}{dx}$ and $\frac{d[f(z)]}{dz}$ are introduced in a single equation. The arbitrary constant introduced in the assumed relation between z and x is then eliminated, and the equation reduced to such a form that one member is apparently a function of z, and the other of x. This last process we call the *separation of the variables*.

As x and z may denote any two values of the independent variable, the apparent functions mentioned above will necessarily be identical in form, and (since they constitute the two members of an equation) identical also in value. This value will be constant, since either member of the equation is a functional expression, which does not change its value with x.

The determination of this constant is then effected by the differentiation of some algebraic identity.

17. The following method of deducing the differential of the product from that of the square is substantially the same as the one employed in *Vince's Fluxions.*

THE DIFFERENTIAL OF THE PRODUCT.

If x and y denote any two variables, xy is a function of both, and its differential depends upon x, y, dx, and dy.

In order to derive this differential, we express xy by means of squares, since we have already obtained a formula for the differentiation of the square. Thus, from the identity

$$(x + y)^2 = x^2 + 2xy + y^2,$$

we derive $\qquad xy = \tfrac{1}{2}(x + y)^2 - \tfrac{1}{2}x^2 - \tfrac{1}{2}y^2.$

Differentiating, $\quad d(xy) = (x + y)(dx + dy) - xdx - ydy,$

or $\qquad\qquad d(xy) = ydx + xdy.$

From the differential of the product, that of the quotient and power are readily deduced by the ordinary methods. We pass now to Transcendental Functions.

TRANSCENDENTAL FUNCTIONS.

18. To deduce the differential of the logarithmic function, we employ the method exemplified in the case of the square.

The symbol $\log x$ is here used to denote the logarithm of x to any base, and $\log_b x$ is used when we wish to designate a particular base b.

Let $z = mx$, then will $dz = mdx,\dots\dots\dots(1)$

$\log z = \log m + \log x$, $\therefore$ $d(\log z) = d(\log x)\dots\dots(2)$

Dividing (2) by (1),

$$\frac{d(\log z)}{dz} = \frac{d(\log x)}{mdx},$$

substituting for m its value $\frac{z}{x}$, and separating the variables, we obtain

$$z\frac{d(\log z)}{dz} = x\frac{d(\log x)}{dx} \dots\dots\dots(3)$$

This equation is true for all values of x, z, dx, and dz, and may be written in the form

$$zf'(z) = xf'(x);\dots\dots\dots\dots(3')$$

moreover, the values of x and z are entirely independent, because their assumed ratio m has been eliminated. Now, since the form of the first member of equation (3) or (3') is the same as that of the second member, this equation shows that the value of the expression $x\frac{d(\log x)}{dx}$ is unchanged when x is changed to z, the latter denoting *any other* value of the independent variable, that is, the expression is independent of the value of x ; it will be found, however, to depend upon the base of the system of logarithms to which $\log x$ belongs.

Denoting now the base by b, we put

$$x \frac{d(\log_b x)}{dx} = B \dots\dots\dots\dots\dots(4)$$

In this equation, B can be dependent upon no quantity except b. Equation (4) may be written in the form

$$d(\log_b x) = \frac{Bdx}{x}; \dots\dots\dots\dots\dots\dots(5)$$

similarly, if a is the base, we use A to denote the constant value of the expression in equation (3), thus—

$$d(\log_a x) = \frac{Adx}{x} \dots\dots\dots\dots\dots(6)$$

A relation between A and B is found by differentiating, by (5) and (6), the identical equation

$$\log_a x = \log_a b \, \log_b x,^* \dots\dots\dots\dots(7)$$

and thus obtaining,

$$\frac{Adx}{x} = \log_a b \frac{Bdx}{x} ;$$

whence

$$A = B \log_a b = \log_a b^B ;$$

therefore, by the definition of a logarithm,

$$a^A = b^B \dots\dots\dots\dots\dots\dots\dots(8)$$

Now, it is obvious that the value of a^A cannot depend upon

---

* This identity is most readily obtained thus,—by definition

$$x = b^{\log_b x},$$

taking the logarithm to the base a of each member, we have

$$\log_a x = \log_b x \, \log_a b.$$

b, hence equation (8) shows that the value of b^B likewise cannot depend upon b; b^B must, therefore, have a value entirely
independent of the base of the system of logarithms to which
$\log x$ belongs.   Denoting this constant value by e, we write

$$b^B = e \dots\dots\dots\dots\dots\dots\dots\dots\dots (9)$$

Taking the logarithm of each member of equation (9), adopt-
ing this constant as a base, we have

$$B \log_e b = 1,$$

whence

$$B = \frac{1}{\log_e b} \dots\dots\dots\dots\dots\dots\dots (10)$$

Introducing this value of B in equation (5), we obtain

$$d(\log_b x) = \frac{dx}{\log_e b \,.\, x}.$$

If the logarithms are taken in the system whose base is e,
the last equation reduces to

$$d(\log_e x) = \frac{dx}{x} \dots\dots\dots\dots\dots\dots\dots (11)$$

19. The constant e is the base of the Napierian system of
logarithms.   In Art. 22, it is shown that this constant is
greater than 2, and less than 3: the more exact determination
of its value is however deferred until the student is able to deduce the requisite series by means of Maclaurin's Theorem.

Articles 20, and 21, upon which Art. 22 depends, will serve
incidentally to show how the method of rates is applied to
finding the formula for the differential of an area.

20. The equation of the hyperbola referred to its asymptotes
is

$$xy = c^2;$$

taking $c=1$, and the axes rectangular, we have for the equation of the rectangular hyperbola in the figure,

$$xy=1, \qquad \text{or} \qquad y=\frac{1}{x}.$$

Draw the ordinate of the vertex (1, 1), and suppose a variable ordinate to move from this position toward the right. The area generated by this ordinate will be an increasing variable quantity, which may be regarded as a function of x, and denoted by $F(x)$.

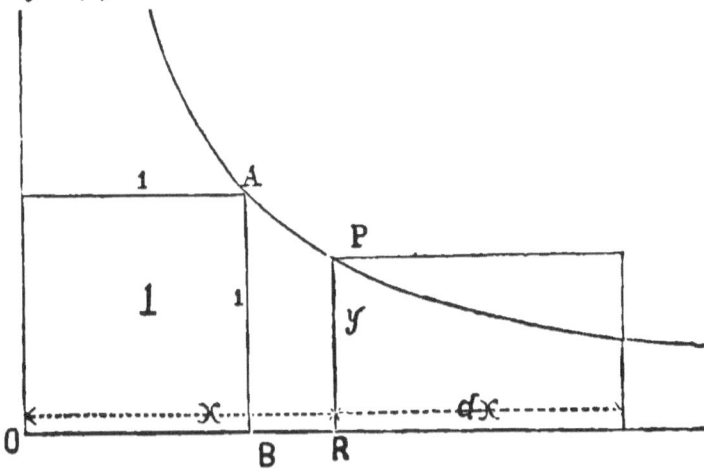

At the instant x has attained a given value, the corresponding rate of $F(x)$ may be measured by giving the area a continuously uniform rate identical with that which it has at that instant [see Art. 8]. This is effected by supposing y to remain constant, and x to increase uniformly from and after the given instant with the rate dx, thus rendering the rate of the area uniform. In the figure, the area $A\,B\,R\,P$ represents the value of $F(x)$ at the given instant, and the rectangle ydx its rate at the same instant. Hence we have at every instant

$$d\{F(x)\}=d(\text{Area})=ydx.$$

This equation, it may be remarked, has been derived without regard to the relation between y and x, and is therefore applicable, whatever be the equation of the curve.

21. In the present case, since $y = \dfrac{1}{x}$, we have

$$d\{F(x)\} = \frac{dx}{x} = d(\log_e x),$$

by equation (11) of Art. 18. Thus $F(x)$, the expression for the area, varies at the same rate as the Napierian logarithm of x.

Now, when $x = 1$, the area reduces to zero, likewise $\log_e 1 = 0$, or

$$F(1) = \log_e 1;$$

therefore, $F(x)$ and $\log_e x$, starting from the same value, and constantly increasing at the same rate, we have

$$F(x) = \log_e x.$$

That is, the area included between the hyperbola, the axis of x, and the ordinates corresponding to x and 1 respectively, is equal to the Napierian logarithm of x. It is easily inferred by subtraction that *the area included between any two ordinates is the logarithm of the ratio of the abscissas.*

22. Let the ordinates corresponding to the abscissas 2 and 3 be drawn; then the area included between the ordinates corresponding to 1 and 2 will represent $\log_e 2$, and that included between those corresponding to 1 and 3 will represent $\log_e 3$. Completing the square $AD\,2\,1$, we see that $\log_e 2$ is less than one.

Draw a tangent to the curve at the point $(2, \tfrac{1}{2})$; this tangent, the axis of x, and the ordinates corresponding to 1 and 3 form a trapezoid of which the sum of the parallel sides is

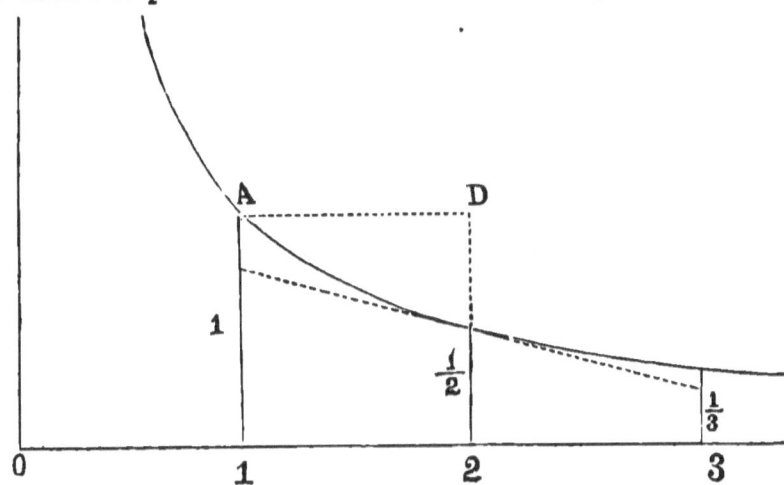

one, and the distance between them two; its area is therefore one, and the area representing $\log_e 3$ is greater than one. Therefore

$$\log_e 3 > 1 > \log_e 2.$$

Now, since $\log_e e = 1$, this inequality is equivalent to

$$\log_e 3 > \log_e e > \log_e 2$$

or

$$3 > e > 2.$$

The Napierian base is therefore intermediate in value between 2 and 3.

The differentials of exponential functions of the forms a^x and x^v are easily derived from the formula for differentiating the logarithmic function.

THE DIFFERENTIALS OF THE SINE AND THE COSINE.

23. Let the variable angle θ be generated by the rotation of the radius a about the origin of rectangular co-ordinates start-

ing from the position OA. The extremity P of the radius moves in the circle,

$$x^2 + y^2 = a^2 \dots\dots\dots\dots\dots\dots\dots(1)$$

and generates the variable arc s.

Let PB and BP' represent the rates respectively of the abscissa and ordinate of P when it arrives at the position indicated in the figure. Then will PP' be a tangent as in Art. 12; and, moreover, the length PP' will represent the actual velocity

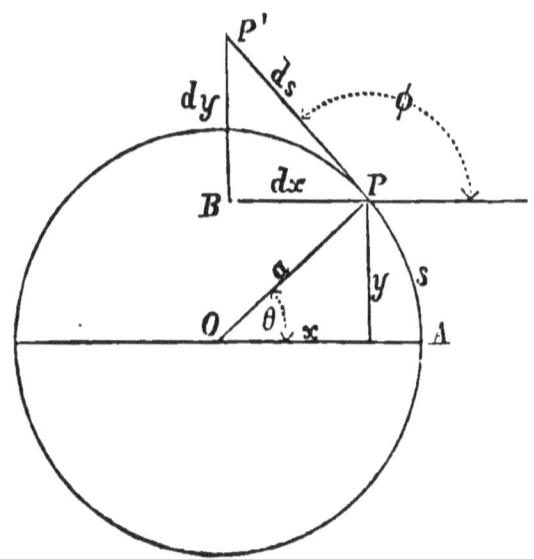

ds, of the point P, because it is the space through which P would move in the unit of time, were dx, dy, and consequently ds, to become constant.

Now, $\qquad \sin\theta = \dfrac{y}{a}$, and $\cos\theta = \dfrac{x}{a}$ $,\dots\dots\dots(2)$

$\therefore \qquad d(\sin\theta) = \dfrac{dy}{a}$, and $d(\cos\theta) = \dfrac{dx}{a}\dots\dots\dots(3)$

In equations (3) we have to express dy and dx in terms of θ and $d\theta$.

Denoting by ϕ the inclination to the axis of x of the direction of the motion of the generating point when s is *increasing*, we have

$$dy = \sin\phi.ds, \quad \text{and} \quad dx = \cos\phi.ds,\ldots\ldots\ldots(4)$$

and, substituting in equations (3), we obtain

$$d(\sin)\theta = \sin\phi.\frac{ds}{a}, \quad \text{and} \quad d(\cos\theta) = \cos\phi.\frac{ds}{a} \ldots\ldots\ldots(5)$$

In the figure, ϕ being in the second quadrant, and ds being positive, dx is negative.

Differentiating equation (1),

$$xdx + ydy = 0,$$

whence

$$\tan\phi = \frac{dy}{dx} = -\frac{x}{y};\ldots\ldots\ldots\ldots\ldots (6)$$

or, since

$$\frac{y}{x} = \tan\theta,\ldots\ldots\ldots\ldots \ldots\ldots (7)$$

$$\tan\phi = -\cot\theta = \tan(\theta \pm \tfrac{1}{2}\pi);$$

therefore

$$\phi = \theta + \tfrac{1}{2}\pi\ldots\ldots \ldots\ldots\ldots\ldots(8)$$

In equation (8) we take $\theta + \tfrac{1}{2}\pi$, because ϕ has been defined as the angle between the positive directions of ds and dx.

Now, in equations (5), we substitute

$$\sin\phi = \sin(\theta + \tfrac{1}{2}\pi) = \cos\theta,$$

$$\cos\phi = \cos(\theta + \tfrac{1}{2}\pi) = -\sin\theta,$$

and

$$-\frac{ds}{a} = d\theta, \quad \text{since} \quad \frac{s}{a} = 0.$$

Whence

$$d(\sin\theta) = \cos\theta d\theta,$$

and

$$d(\cos\theta) = -\sin\theta d\theta.$$

24. Equation (8) shows that, in the case of the circle, the tangent, according to the general definition of Art. 12, is perpendicular to the radius through the point of contact, and is therefore identical with the tangent as defined in Geometry.

The differentials of the remaining circular functions may be deduced from the above formulas in the usual manner.

----

25. In conclusion, we will briefly indicate the mode of adapting this method to applications of the Calculus.

The rate method is, by its definitions, peculiarly adapted to Kinematical and Physical Investigations. (See extract from *Cournot*, Art. 7 of this paper). It is noticeable that in several of the best recent treatises on Dynamics, Newton's Fluxional notation as applied to functions of the time, has been revived, and is freely employed.

26. In the Integral Calculus, as well as in the Differential, the rate method excels in the clearness it gives to the fundamental definitions.

The Indefinite Integral is defined as such a function of the independent variable as will vary with the rate expressed under the integral sign. *A Particular Integral* is any one such function. It is then easily shown that the particular integrals included in an indefinite integral can differ only by a constant. *A Definite Integral* is the amount by which a function, having the rate expressed under the integral sign, actually changes, while the independent variable passes from one limit to the other. Accordingly, an indefinite integral becomes "particular" when the lower limit is fixed, the upper limit remaining variable.

27. In deriving differential expressions to be integrated, a reference to the pages of *Montucla*, cited in the introduction to this paper, Art. 4, will show that the methods employed in works on Fluxions, for deducing the formulas for the rates of areas* and volumes were in general both clear and satisfactory. The underlying principle is that of rendering a non-uniform rate measurable by removing the cause of its variable character (in this case, the cause is the variation of the generating line or area).

In deriving other formulas, like those for moments and pressures, the methods employed in these treatises were not so satisfactory, and at this point we propose to show that *the comparative rates of two quantities can be obtained by taking the ratio of simultaneous increments and passing to the limit.*

It is thought that the introduction of limits at this stage of his progress will present little difficulty to the student, now familiar with the notion of continuity; while it will prepare

---

* See also Article 20 of this paper.

him for the study of writers who employ limits and infinitesimals in the applications of the Calculus; and serve to
demonstrate the fundamental identity of the different methods
of treating this important branch of mathematics.

ON

A NEW METHOD

OF OBTAINING THE

DIFFERENTIALS OF FUNCTIONS

WITH ESPECIAL REFERENCE TO THE

NEWTONIAN CONCEPTION OF RATES OR VELOCITIES

BY

. J. MINOT RICE

PROFESSOR OF MATHEMATICS IN THE UNITED STATES NAVY

AND

W. WOOLSEY JOHNSON

PROFESSOR OF MATHEMATICS IN SAINT JOHN'S COLLEGE ANNAPOLIS MARYLAND

REVISED EDITION

NEW YORK

D. VAN NOSTRAND PUBLISHER

23 MURRAY STREET AND 27 WARREN STREET

1875